Improve the Speed of Your Pinewood Derby Using a RobotBASIC Timer

Copyright June 2014 by John Blankenship

The Cub Scout's Pinewood Derby® is an exciting contest for everyone. Scouts love trying to build a better and faster car than everyone else. If you are really into the Derby, you probably know that a few simple searches on Yahoo or Google will give you *many* ways to improve your car. Figure 1 shows a basic kit for building a Derby Car. As you can see, it is little more than a block of wood ready to become whatever you imagine it to be.

![Figure 1 photo of Pinewood Derby Kit]

Figure 1: A Pinewood Derby Kit invites your creativity.

In general, the potential speed improvements for your car will fall into two categories. First, it is imperative that gravity creates the most momentum possible. Second, the

1

resistance against your car's movement must be minimized. This resistance can come from wheel friction, aerodynamics, etc.

Improving Your Car's Speed

Increasing the weight of your car can greatly improve its momentum, but properly placing that weight is essential. Smoothing the nails used as axels and having them properly aligned is also important. If you search the Internet you will find many opinions and suggestions on topics such as these.

Just knowing the kinds of things that will improve your car's performance is not enough though. If you truly want to maximize your car's speed, you need a way to determine IF the things you are doing are actually making a difference. Let's look at some examples.

Suppose you want to add weight to your car, but you don't know exactly where to put it. From what you have read on the Internet, you know it is better for the weight to be on the rear of the car, but might it be better in front of the rear axel or behind it? Would it work better if the weight was higher on the back of the car or would a lower position produce better results?

What about preparing the nails used to mount your wheels. Even slight imperfections and ridges can impair your cars performance. Your web searches will provide you lots of ideas on what to do, but how do you know if what you are doing is working? How do you know if you have polished your axels enough?

But, How Do You Know

The answer to all these questions is easy. What you need is a way to accurately measure the amount of time it takes your car to complete some length of the track. Obviously, it would be better if your Troup made the actual track available at various times for testing purposes, but you can test the changes you make using a simple homemade ramp. In general, if the changes you make increases your car's speed on a makeshift track that you build, it will do better on other tracks as well.

Timing your car's speed is not easy. Trying to do it manually, with a stop watch, is unlikely to produce rewarding results because you will always make small errors when pushing the start and stop buttons on the watch. What is really needed is a way for a computer to time the car for you.

Writing a computer program to time your car is VERY easy. Attaching sensors to your computer to observe your car's movement though, can be more difficult and in many cases it can be expensive. One of the goals of this project is to keep the cost as low as possible. This will force you to learn a few things and do a little work, but that can be the fun part if you like building things – and if you are constructing a Derby car, you probably love building things.

In order to write a program to time your car we need a language the computer understands. This is not all that different from talking to someone from China or Italy. If you want them to understand what you are asking them to do, you must speak to them in a language they understand.

Programming a computer is just a way of telling it what you want it to do, and you must use a language that it understands. Many computer languages are very cryptic and difficult to use. They can also very expensive.

RobotBASIC
For this project, we will be using a language called RobotBASIC. It is a very powerful language (you can use it for many other things too), but it is very easy to use. And, it is FREE. It was designed to run on a PC, so that is the type of computer you must use if you want it to be able to understand the program. If you use an Apple or other type of computer, you can use the techniques discussed here, but you will have to create your own programs with a language designed for the computer you will use. We will talk more about the program for this project shortly.

Normally when a computer is going to interact with the outside world, it needs special electronic circuitry that makes it possible for events in the real world to be monitored or controlled by the computer. In this case we want to monitor when your car starts down the track, and when it reaches the finish line. We will use that information to determine how long it took the car to complete the track.

We could buy special hardware, often referred to as Input/Out Ports, that could help perform this monitoring for us, and if we wanted to monitor a lot of things (like we might want to do if we were building a robot, for example) then such equipment would generally be a good idea. RobotBASIC and many other companies sell such parts, should you ever want to build a robot or other complex computer-based project.

Monitoring the car is a relatively simple task so we really don't need a lot of complex hardware. As mentioned earlier, we only need to know when the car starts, and when it crosses a finish line. If you think about your computer's mouse, it has the capability of monitoring two things – because it has two mouse buttons. Some mice have more than two buttons, but for our purpose, two is enough. We will be using the LEFT and RIGHT mouse buttons to time the car.

As we will see later, it is very easy to create a small RobotBASIC program to watch the mouse buttons and start a timer when the left mouse button is pressed (or perhaps released). The program will stop the timer when the right button is pressed and display the total time between the button presses. This will provide a very accurate reading of the time between the pressing of the two mouse buttons.

What we need, of course is for your car to be able to press the mouse buttons all by itself. This could be done in many ways. You could, for example, just mount your computer's mouse at the top of your racetrack ramp and push your car against the left mouse button. When you release the car, it will start down the ramp – but, as it starts to move, it will also release the left mouse button and we can have the program detect that action and start a timer. If we only had a way to stop the timer when the car reached the finish line, we would be ready to program. Unfortunately, both of the mouse buttons are on the mouse itself and we need the second button to be near the end of the track. As long as we are going to create a remote button at the end of the track we might want to create one for the beginning too, as it could be easier to use than pressing the car against the mouse itself.

The point is that we need a way to effectively place the right mouse button at the finish line where the car could activate it as it passes by. This sounds impossible, since the mouse, in the example situation just described, is mounted at the top of the ramp – but it is not as impossible as it sounds.

The plastic buttons on a mouse are typically mounted over a very small switch inside the mouse. When you press on the plastic button it presses on a lever or contact point on a switch inside the mouse, causing the switch to close. When this switch opens and closes the electronics in the mouse sends singles to your computer that can be read by programming languages like RobotBASIC. The details of how all this happens is complex but the good news is that you don't have to understand the hard stuff to modify your mouse so that it can be used to create a Pinewood Derby Timer.

Before we talk about modifying your mouse, let's discuss switches in general. A switch is nothing more than two conductors (pieces of metal) that either touch each other or not, based on a lever or button used to turn the switch on (touching) or off (not touching). The electronics symbol for a switch is shown in Figure 2. The two circles represent the connection terminals and the line with an arrowhead on it represents the connection lever. It is shown here in the open (or off) position

Figure 2: This is the electronic symbol for a switch.

Figure 3 shows the same switch, but in both of its possible states. The top view is the same as Figure 2, which shows the switch in its OFF position. When the switch is closed (the bottom view) electricity can flow through the switch. Engineers call these SPST switches. SPST stands for Single Pole Single Throw because such switches have only one contact arm, and one ON position.

Figure 3: A switch can be ON (bottom) or OFF (top)

Figure 4 shows a simple way to use a switch. The drawing is how engineers represent various parts. On the left is the symbol for a battery. The symbol on the right is a lamp or light bulb. The lines connecting all the parts represent wires. If you close the switch, the current will flow from the battery through the bulb and back to the battery. You can think of the battery as a pump that forces electrons through the lamp (causing it to light) and back to the battery. The switch acts as a valve that lets the electricity flow when it is turned on.

Figure 4: When the switch is ON, the battery lights the lamp.

Figure 5 shows some pictures of real switches. The switch on the right uses a lever to turn it ON or OFF. Notice the two terminals on the bottom of the switch for connecting wires. The switches on the left are push button switches. To turn them ON, you must hold down the button. Releasing the button turns them OFF. These switches have terminals on their bottom side too. Notice that one of the switches has wires soldered to its terminals to make it easy to connect to other devices. The switch on the right is called a toggle switch. It uses a lever to toggle between the ON and OFF state.

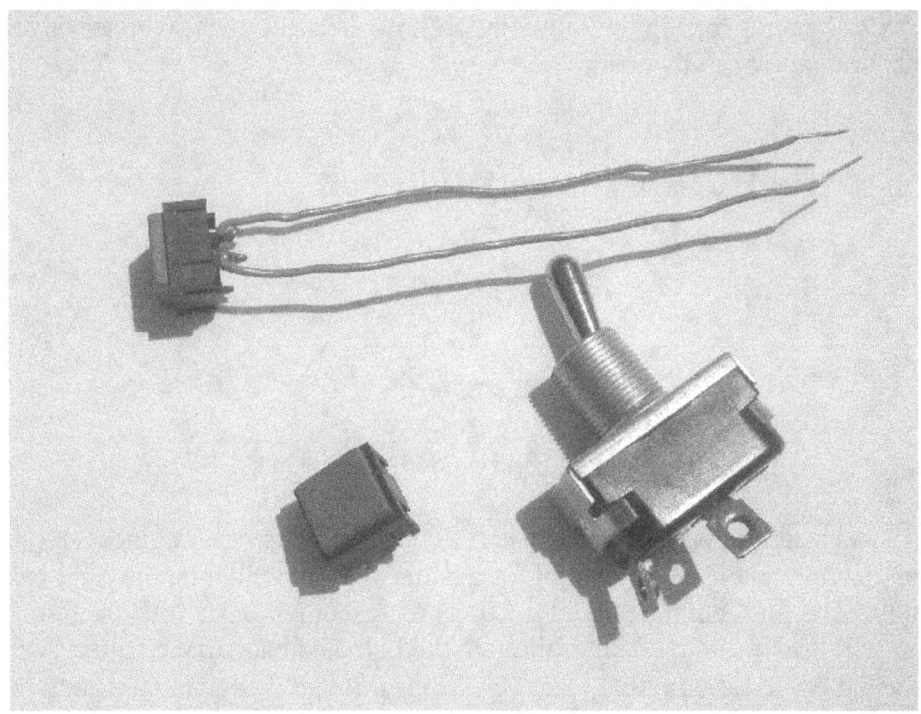

Figure 5: These real switches have two terminals.

Some switches have three terminals instead of two as shown by the symbol in Figure 6. The connecting lever either connects between the left terminal and the top right-hand terminal, or between the left terminal and bottom right-hand terminal. You can see this better in Figure 7.

Figure 6: Some switches have 3 terminals.

The top and bottom views of Figure 7 show the switch in two positions. Whether you consider the positions ON or OFF depends on how you connect the switch. If you just use the two bottom terminals, then it works just like the two-terminal switch shown in Figure 2. Three-terminal switches like this are called SPDT (Single Pole Double Throw) because they have two contact positions (double throw)

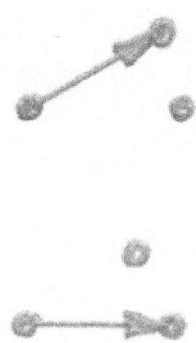

Figure 7: Three terminal switches have more options.

Figure 8 shows some three-terminal switches. The one on the left is a slide switch. When the slide button is on the left, as shown, the center terminal is connected to the left terminal. If the slider is pushed to the right, the center terminal is connected to the right terminal. If you only use the center terminal and one other, then this switch acts just like two terminal switch. The OFF position depends on which terminals you use.

Figure 8: Two real 3-terminal switches.

The switch on the right of Figure 8 also has three terminals. Typically the common terminal is on one end of toggle switches like this one, but you cannot know for sure without testing. When the lever is moved to the left and right, the common terminal connects to one or the other of the remaining terminals. Sometimes switches like this have a center position for the toggle-lever that allows the common terminal to NOT connect to either of the other two terminals.

Figure 9 shows a simple example for how a 3-terminal switch can be used. When the switch is in the position shown, the lamp L2 is ON. If the switch is moved to the lower position, then lamp L1 will light. Often the two terminals are called normally-open (NO) or normally-closed (NC) to indicate whether the terminal is ON (closed) or OFF (open) when the switch is in its *normal* position. Normal can sometimes be a relative term, but not always. For example, a push-button switch would be considered to be in its normal state when it is NOT being pressed. As another example, some toggle switches have an internal spring that keeps them in a normal position unless the lever is pushed (much like a push-button switch). When switches do not physically have a normal position, you must decide how to mount the switch and which terminals to use to meet your needs.

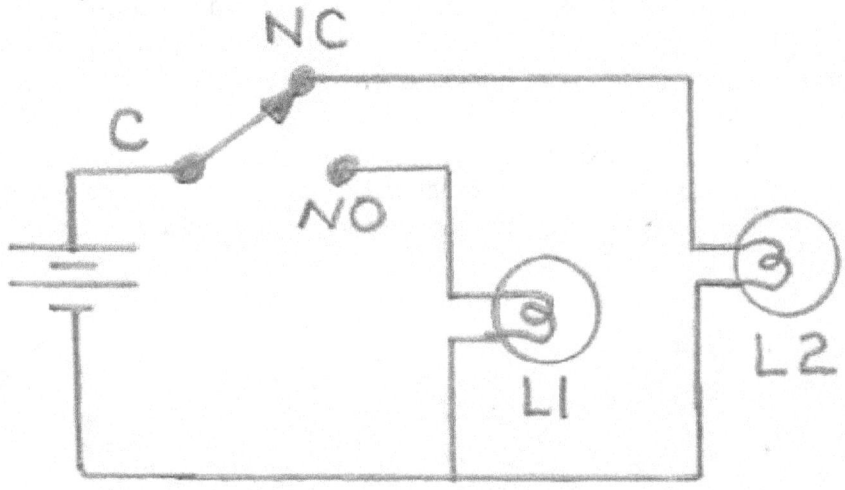

Figure 9: In this example, the switch can turn on either of two lights.

You could find out which terminal is connected to the common terminal for each switch position by connecting it with a battery and a lamp as shown in previous figures. Another way is to use a meter, as shown in Figure 10. Meters like this can be purchased very inexpensively. I purchased this one from Harbor Freight for about $10.

Meters have MANY uses such as measuring volts, amps and ohms. There are many Internet sources to help you understand how to use them, but for our purposes, the meter will only be used to indicate if two terminals are connected together. We can do that by measuring resistance (ohms) on the lowest scale. The meter in Figure 10 is set to the lowest resistance scale and the leads (which are used for measuring) are not together – thus an open circuit (like a switch that is open, or OFF). Under these circumstances the resistance is much too high to measure, so the meter must indicate an open circuit. This particular meter gives a reading of 1, with blank space between it and the decimal point, to indicate an open circuit.

Figure 10: Meters can be used to determine the functions of the terminals of a switch.

If the meter leads are connected together with a clip lead as shown in Figure 10, then the meter should read something very close to zero ohms to indicate a short circuit. Notice the meter in Figure 11. It reads 4.6 ohms. The actual resistance of the clip lead is probably even smaller than this, but we have to remember this is a very inexpensive meter. The point of this is that the meter allows us to determine if its two leads are connected together. If you connect the leads to two terminals on a switch, a open-circuit reading indicates the terminals are not connected while a low reading indicates they are connected.

Let's summarize this for clarity. If we connect the leads to the terminals of a 2-terminal switch. When the switch is ON, the meter should read something very close to zero. When the switch is OFF, the reading will be as shown in Figure 9 or something similar based on the particular meter you are using. You can find the two readings to look for with your meter, by simply shorting the meter leads together and taking them apart.

Once you know what readings to look for, you can connect the leads to two terminals of a 3-terminal switch. If the readings change when you turn the switch ON and OFF, then you can determine which terminals are connected together for different switch positions. If the meter always reads open, regardless of whether the switch is ON or OFF, then the meter leads must be connected to the two main switch terminals and the unused terminal is the common terminal. Refer back to Figure 7 to better understand this. If the meter is connected to the two right-hand terminals, then it will never see them connect regardless of the switch position.

If you use a meter (or even a battery and a flashlight light bulb) you should be able to determine which terminals are connected together when the switch is in either of its two possible positions.

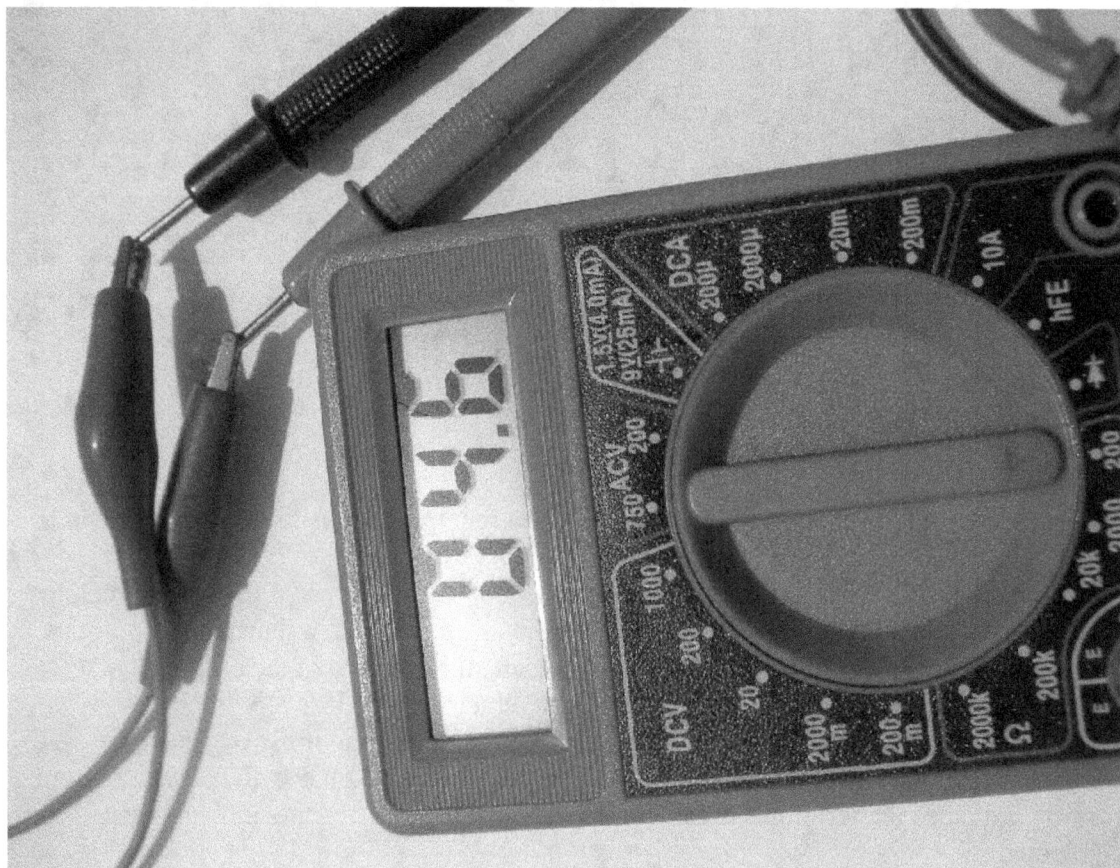

Figure 11: This meter is measuring 4.6 ohm.

We want our car to be able to press one of two switches for us, depending on where the car is on the track. The first switch will be at the start of the track and the other near the end. All of the switches we have seen so far are far too hard to press to be used to detect the car. Figure 12 shows a different kind of switch that is perfect for our purposes.

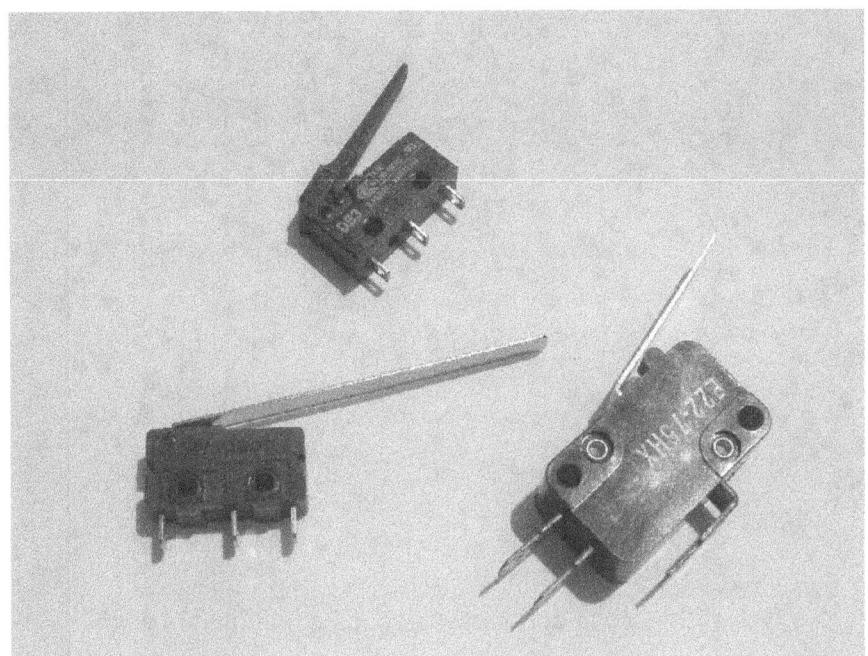

Figure 12: Three snap-action switches.

Notice that each of the switches in Figure 12 has a long lever to make it easy to activate the switch with your car (more on this later). Such switches are often called snap-action. They are typically used in commercial products to detect things like whether a refrigerator door is closed. Most snap-action switches are 3-terminal switches, like all the switches in Figure 12.

We must mount the switches so that the car will turn them ON when it is over them. You could cut a slot in the ramp to mount the top switch as shown in Figure 13. If you do not have room to mount the switch this way at the end of the track, you could build a small ½ inch high ramp at the finish line. When the car rolls onto the ramp and over the switch (see Figure 14), the switch will turn ON.

As mentioned, we must have one switch at the top of the ramp and the other near the end of the track, as shown in Figure 15. Now that we understand switches, we need a way for the computer to know whether each switch is ON or OFF.

RobotBASIC can easily detect when the left and right buttons on a mouse are pressed. These buttons are just switches that cause changes in the circuit that the computer can detect. If we connect the switches on our track in parallel with the mouse switches, then the car will be able to effectively press the mouse buttons.

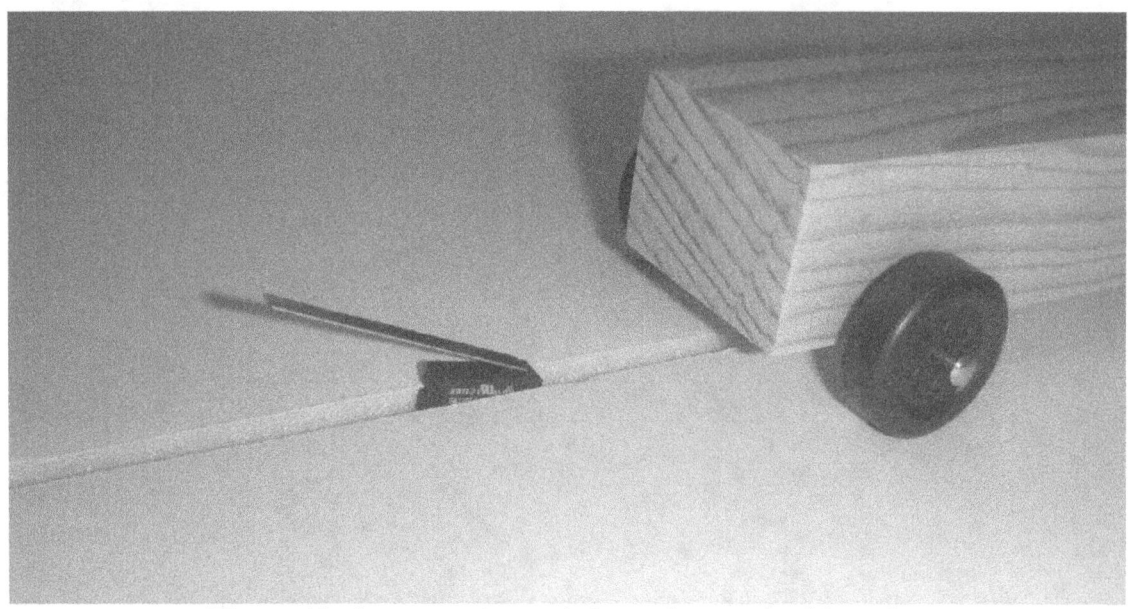

Figure 13: A snap-action switch can be mounted in a slot.

Figure 14: The switch will come ON when the car rolls over it.

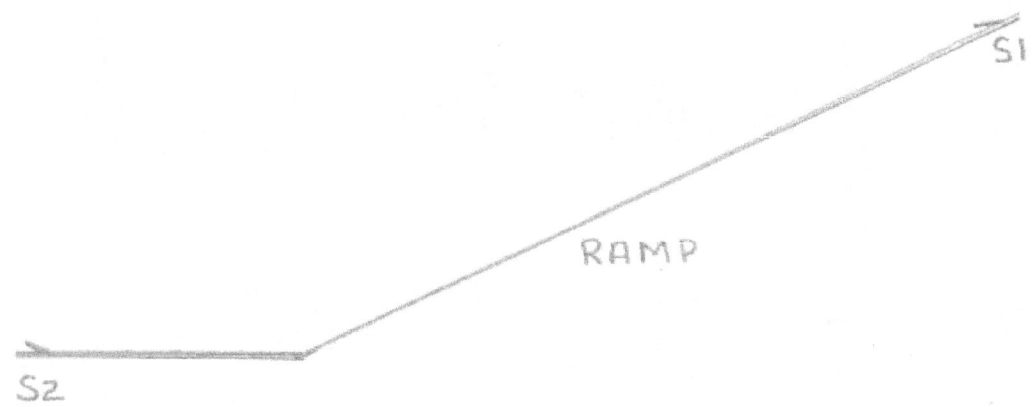

Figure 15: Switch 1 is at the top of the ramp and Switch 2 is near the end of the track.

Figure 16 shows the basic idea of how to connect a remote switch in parallel with an existing switch. Remember the circuit shown in Figure 4. It turned on the lamp when the switch was turned ON. In Figure 16 we have that same circuit, but have connected a new switch (S1) in parallel with the original switch (S2). Now, if you press either S1 or S2, the lamp will light.

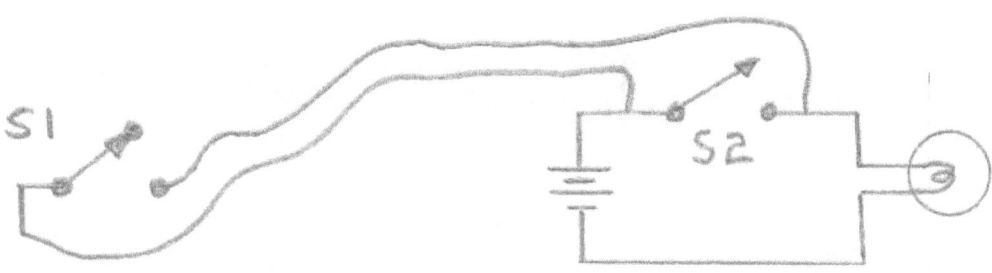

Figure 16: A remote switch can be used to control the lamp.

The switches in a mouse are typically connected as shown in Figure 17. Both the one for the Left Button and the one for the Right Button are connected together on one end of the switches. The three wires extending from the two switches will connect to various components inside the mouse (perhaps resistors, transistors, etc). We know that the electronic circuit in the mouse enables the computer to determine the state of either of these switches and that is all that is important.

Figure 18 shows how we will connect the two remote ramp switches (Figure 15) in parallel (as we did in Figure 16) with the mouse switches. The wires connecting to the mouse will have to be long enough to reach the switches on the ramp.

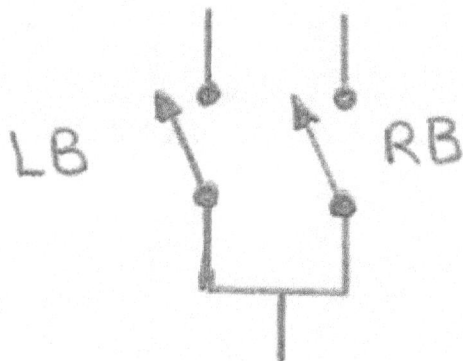

Figure 17: The switches inside a mouse typically have two terminals connected together.

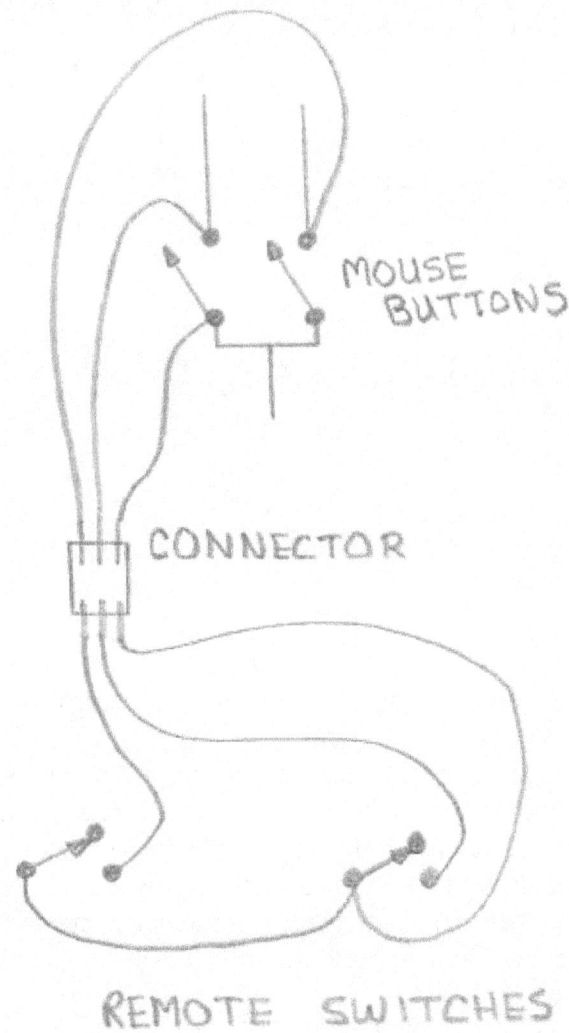

Figure 18: The track switches can be connected in parallel with the mouse switches.

Of course, we must open up the mouse and look inside to find the terminals for the mouse-button switches. Obviously, you should not do this with the mouse you normally use. After all, if you mess up the modifications, you could damage your mouse. Luckily, finding an old mouse to use is usually relatively easy. You or a friend might even have a mouse for an old computer you have thrown out. They can often be found at yard sales for almost nothing. Even a new one is not that expensive anymore.

Choosing a Mouse
You can use a standard mouse that has a cable that can connect to your computer, but if you can find a wireless mouse, then that is even better. Figure 19 shows both a wired mouse and a wireless mouse – note the USB transmitter/receiver for the wireless mouse.

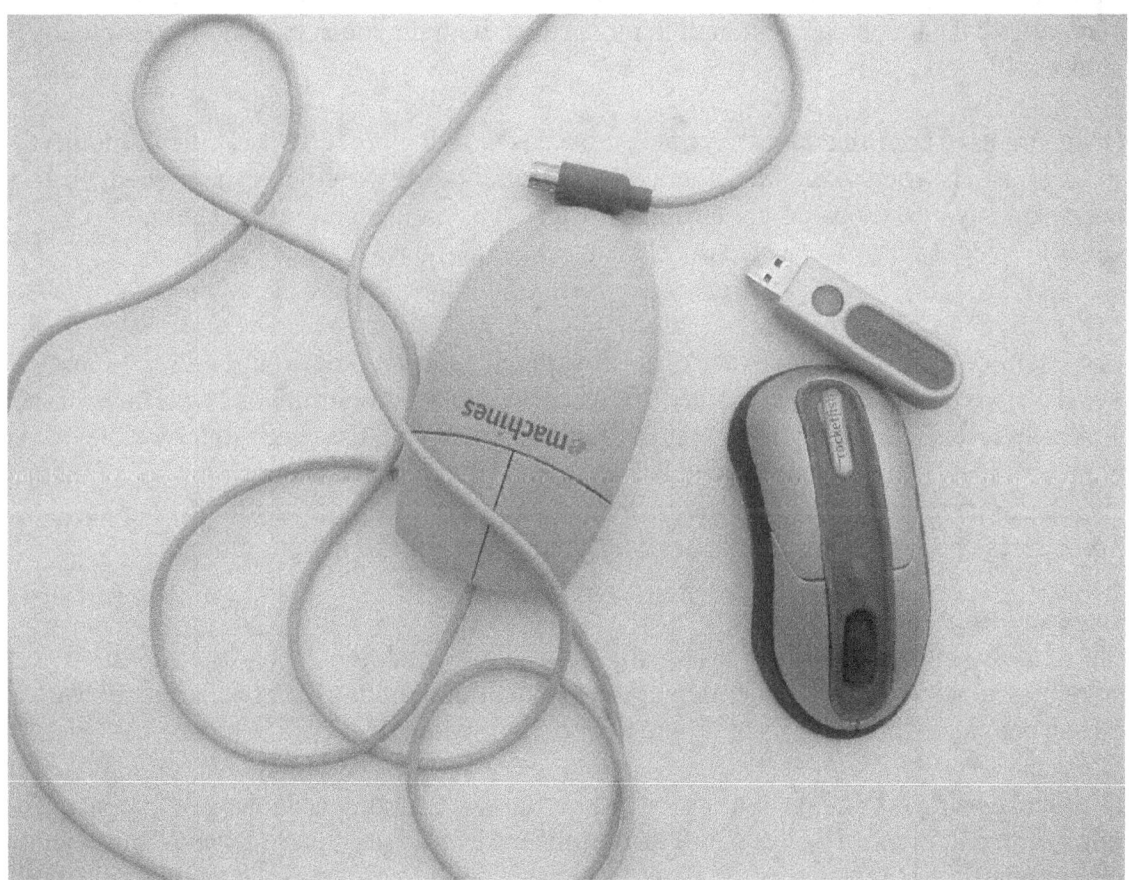

Figure 19: You can use any mouse that can connect to your computer, but a wireless mouse is recommended.

A major advantage of the wireless mouse is that it is not physically connected to your computer. This can sometimes make it easier to connect to your racetrack, but it also means that if you modify your mouse incorrectly, that it cannot damage your computer. As we will soon see, the modifications you must make to your mouse are relatively easy, so it is unlikely that you will have problems if you use a wired mouse for this project. If you don't feel 100% comfortable performing the modifications though, you should use a wireless mouse.

No matter what kind of mouse you use, you should connect it to your computer following the instructions that came with it and make sure it is working properly. On older desktop computers you can just unplug the old mouse and plug in the new one. Older computers had a variety of mouse-connectors, so make sure you find a mouse that will fit the socket on your computer.

Newer mice often plug into a USB port. Most versions of Windows will detect the new mouse and start using it automatically. If your computer is a laptop, it will probably have a touch pad for performing mouse functions. When you add a USB mouse to a laptop, it may disable the touch pad, although in most cases, both the touch pad and the external mouse will work.

Once you have confirmed that your new mouse is working properly it is time to modify it so that it can be used to monitor your car's movements. As previously mentioned, each button on a typical mouse is a 2-terminal switch. The mouse-button switches are generally very small and assume the ON state when a mouse button is pressed.

In most mice, the two switches are connected together as was shown in Figures 17 and 18. The two terminals that are joined are usually connected to *ground*. The term *ground*, generally refers to a common point in a circuit diagram. We now need to locate the two main terminals for your mouse switches and the common ground connection. To find the switches, you must first remove the necessary screws and take the cover off your mouse. Sometimes one or more of the screws is located under a label on your mouse.

Locating the Mouse Switches
Physically, a mouse-button switch usually looks like a small box. You can find the switch by looking directly under one of the plastic mouse buttons. Pressing the mouse button activates the switch itself by pressing on the switch's lever or activator.

In order to help you find the buttons on your mouse, let's look inside the two example mice shown in Figure 19. Let's look first at the wired mouse. Figure 20 shows the circuit board. The left side of the Figure shows the lower portion of the mouse case and the internal circuit board. The right side shows the top of the mouse case.

Notice, in Figure 20, that the two switches are circled on the left side of photo to help you find them. On the right side, the plastic mouse button is also circled. When the top of the mouse is replaced on the lower section, the circled mouse button will be directly over the right mouse switch so that the mouse button can press the tiny switch activator.

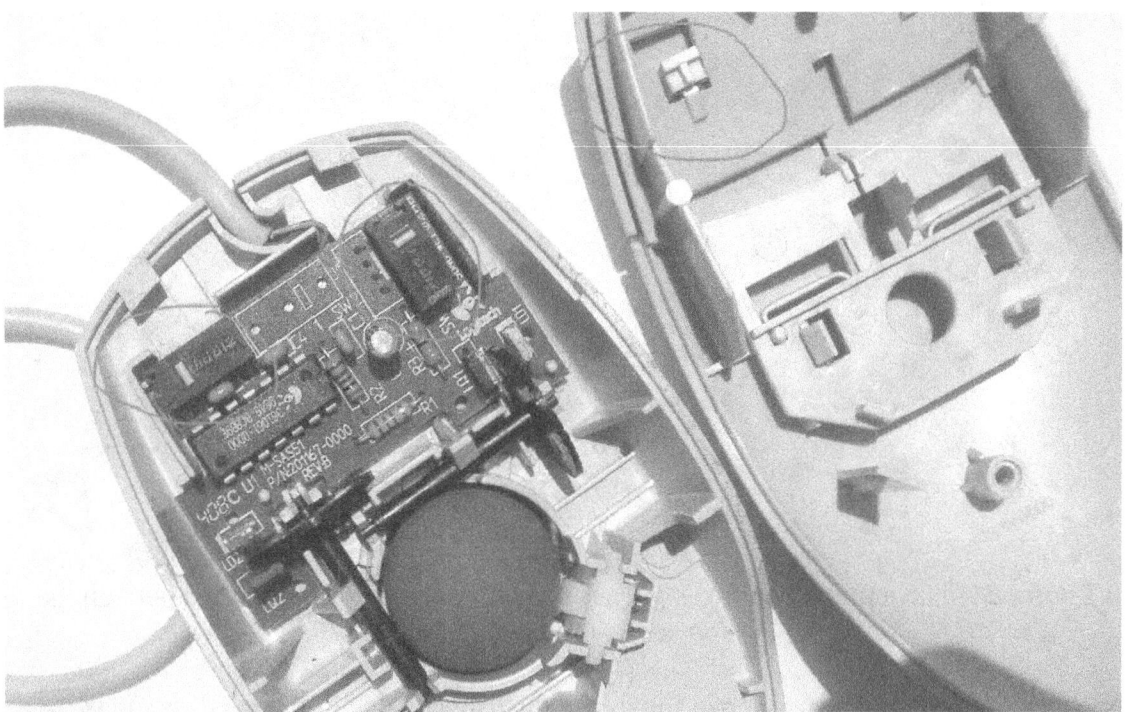

Figure 20: The physical switch is often a rectangular box with a lever or button on its top.

If you turn over the circuit board shown in Figure 20, you will see its bottom side, as shown in Figure 21. The connection points where the two switches are soldered to the circuit board are circled to help you see where they are. Let's look first at the left side of the Figure. Notice that the left terminal is connected to a relatively large area of copper plating – certainly much larger than the plating where the right terminal is soldered. Notice also that the plating for the left terminal runs toward the right side of the board. On close inspection, you will see that this plating also connects to the lower terminal in the right-side circle. This is the common terminal of the two switches (ground). Sometimes it is difficult to see how the common points connect (because some of the connection plating may be on the opposite side of the board). The terminal with the largest amount of plating is almost always the common terminal.

You can use a meter to determine which of the switch terminals are connected together. It should show a low reading when the leads are connected to the two common terminals. You can also connect the meter to the two terminals for each switch. Turning the switch on and off should change the meter readings as described earlier. When measuring resistance with the meter, you should remove the batteries from a wireless mouse or unplug a wired mouse from the computer.

Figure 21: The terminals for the switches shown in Figure 20, are circled here on the bottom side of the circuit board.

Now let's look at another example. Figure 22 shows the inside of the wireless mouse from Figure 19. This is a close-up view, so it only shows one of the two switches. The switch is circled at the bottom of the Figure. The button that presses the switch is circled at the top of the Figure.

Figure 23 shows the reverse side of the circuit board so you can see the terminal points, which are circled for each switch. In both cases, the left terminal in the circle is connected to the largest plated area, identifying it as the probable common or ground terminal. We need to connect wires to the each of the main terminals, and one wire to the common terminal. There is no need to connect wires to both common terminals because they are connected together internally (remember, that is why they are called common).

Modifying the Mouse
Figure 23 shows three wires soldered to the appropriate points. The top wire (yellow if you are reading this in the ebook version which has color photos) is connected to the main terminal of the left mouse switch. The right-most wire in the picture (red) is connected to the main terminal of the right mouse switch. Remember, this is being viewed from the bottom side so the left and right references to mouse switches may seem reversed to you. The final wire (black) is connected to the common terminal of the left mouse switch. You could, of course, have connected it to either of the common terminals.

Figure 22: The switches inside the wireless mouse are similar, but different, than those in the wired mouse.

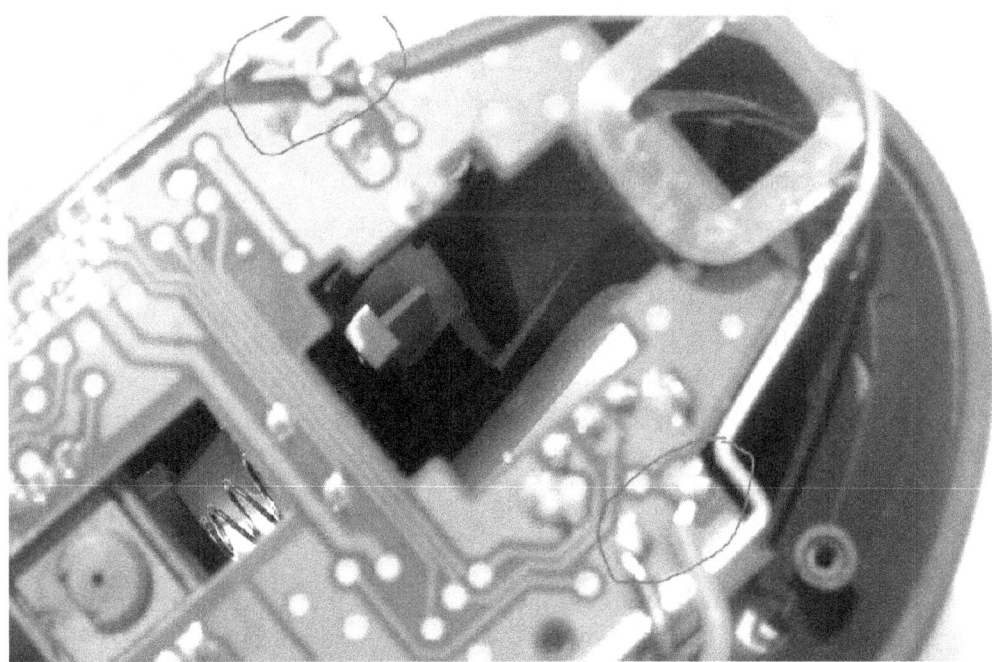

Figure 23: The bottom side of the circuit board (Figure 22) shows the switch terminals with new wires soldered to them.

Figure 24 shows a full view of the back side of the circuit board. The wires leave the mouse through a hole I added for that purpose. Notice the connector at the end of the wires. Using a connector is not necessary but it helps ensure that the wires don't touch each other when not in use. The connector shown here is actually one end of a servomotor extension cable available at most hobby stores. Standard hook up wire can be inserted into the female sockets on the connector and then run to the switches on your racetrack as described in Figure 18. Remember, the connector is just for convenience. If you prefer, you can just use longer wires and run them directly to the remote switches.

Figure 24: This expanded view shows the new wires exiting from the mouse case.

Take care to ensure that the wires do not get in the way when you reassemble the mouse. The final mouse looks like Figure 25. The wires can easily be extended to the remote switches. Note: The common wire has to go to both switches as shown in Figure 19. Remember, the snap-action switches you are using have three terminals, but we will only use two of them. One will be the common terminal. The other should be the normally-open terminal (the one that connects to the common terminal when the switch lever is pressed).

You can test your modified mouse by moving the mouse pointer around your screen and pressing the remote switches to confirm they perform like mouse buttons.

Creating the Program
When you have the remote switches connected to the mouse, we are ready to see how to write a program that can measure how long your car takes to move between the two remote switches.

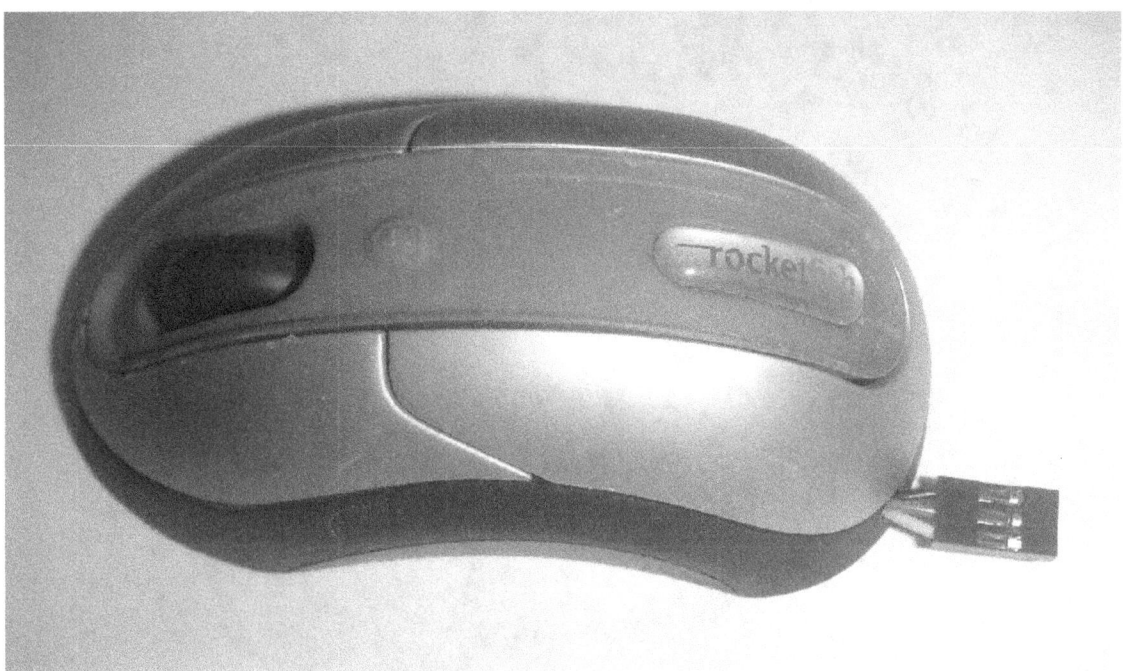

Figure 25: The completed modified mouse.

This programming portion of this project assumes the reader has some basic knowledge of RobotBASIC. If you are totally new to RobotBASIC, it is suggested that you watch some of the many RobotBASIC YouTube videos available from www.RobotBASIC.org (or just search YouTube for RobotBASIC) to give you a little help getting started. Those that need or desire more background information on programming should consider *RobotBASIC Projects for Beginners* and *Robots in the Classroom*. These and other RobotBASIC books (as shown in Figure 26) are available from www.Amazon.com. If you like doing projects like this one, you might especially enjoy the book *Hardware Interfacing with RobotBASIC*. It shows how to connect the computer to external devices like motors and other devices.

RobotBASIC Tutorial
I don't want you to have to purchase other books just for this project though, so here is a short tutorial about programming with RobotBASIC. Who knows, after being exposed to programming, you might discover that you want to learn more about it.

Figure 26: Many books are available to help you learn programming with RobotBASIC.

Starting RobotBASIC

To get started with RobotBASIC, visit www.RobotBASIC.org and visit the FREE
PROGRAM DOWNLOAD tab. Scroll down slightly and click the link ZIPPED DEMOS
& EXE to download a large zip file that includes RobotBASIC, the help file, and many
demo programs. Save the file to a suitable place such as your DOCUMENTS folder and
unzip the file so that you have access to RobotBASIC. Open the new folder and click on
the RobotBASIC EXE to start the program. You may want to create a desktop shortcut
to make starting RobotBASIC easier in the future.

When RobotBASIC executes the first time, it will ask you to review and accept the
license which generally states that RobotBASIC is free, and that you can use it but not
sell it to others. At that point you will see the RobotBASIC editor screen as shown in
Figure 27.

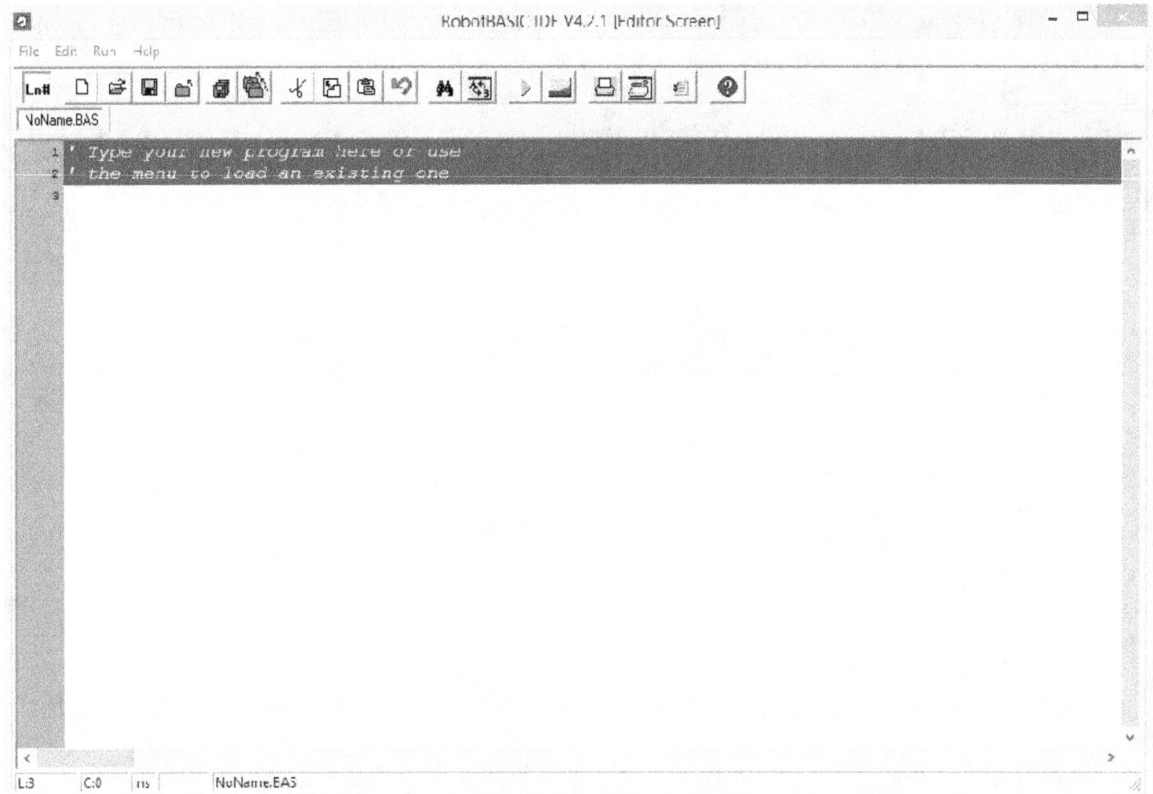

Figure 27: This is the startup editor screen for RobotBASIC

The tool bar at the top of the RobotBASIC editor window offers many shortcuts to items in the main menus. You can, for example, refer to the HELP file by pressing the **?** on the toolbar instead of choosing COMMAND SYNTAX from the HELP menu.

When you write a program you type it into the text area just as you would with any word processor or notepad. You can execute the program by using the menus or pressing the green triangle in the tool bar. For example, type in the short program in Figure 28 and then run it by clicking the green triangle.

```
x=5
y=10
a=x+y-3
print a
end
```

Figure 28: Type this program into the RobotBASIC editor.

When a computer program runs, it executes each statement in order unless control statements cause statements or sections of the program to be repeated or ignored. The program in Figure 28 will print 12 as expected. This printing, as does all output from

RobotBASIC programs, occurs in a new output window (sometimes called the terminal window). Closing the output window returns you to the editor.

If you spell something wrong (**prit** instead of **print**, for example) RobotBASIC will issue an error. Closing the error window will return you to the editor and highlight the offending line so that you can examine it and fix the problem.

Variables, such as **x**, **y**, and **a**, in this example program are case sensitive – that is **a** is not the same as **A**. RobotBASIC commands though (such as **print**) are NOT case sensitive, so **print** is the same as **Print**. You cannot have a variable named the same as command, but you could, for example, have a variable named **MyPrint** or **print1**.

You can save or retrieve (OPEN) programs using the FILE menu or the toolbar icons, again much the same way as you would in a word processor.

A Robot Simulator

One of the great features of RobotBASIC is its integrated Robot Simulator. The simulated robot is easy to use and provides sensory capabilities far beyond that of most educational robots. We won't need the simulator for this project, but let's learn how to control the basic movements of the simulated robot in order to get a little more practice programming. The simulated robot is far more interesting than just adding numbers together.

The simulated robot can be initialized at a specific **x,y** position on RobotBASIC's output or terminal screen with the following statement. Note: The terminal screen is 800 pixels wide and 600 pixels tall. The upper left-hand corner of the screen is the starting point, with coordinates 0,0.

```
rLocate 400,300
```

Notice that the command starts with the letter **r**. All the robot-related commands in RobotBASIC start with an **r**. Type the above command into RobotBASIC's program space (as a new program – erase any existing statements) and run the program by pressing the GREEN triangle near the top of the main RobotBASIC screen. When you do, you will see the simulated robot appear at the center of the output screen as shown in Figure 29.

Figure 29: The robot has been initialized at the center of the output screen.

After the robot has been initialized, it can be moved around the screen as demonstrated by the program in Figure 30. The **rLocate 300,500** will place the robot on the screen at the **x,y** position 300,500. The simulated robot is 40 pixels in diameter, so the **rForward 120** command will move it a distance equal to three times its diameter.

```
rLocate 300,500
rForward 120
rTurn 90
```
Figure 30: This program moves the robot a distance equal to three times its diameter, then turns it 90° to the right.

Consider the idea that the robot will move in a square if the last two commands in Figure 30 are repeated three more times. We could, of course, just type the commands in four times, but computer languages have flow-control structures that allow the creation of loops that can execute a series of commands multiple times. Figure 31, for example, will repeat the forward and turn commands four times, causing the program to move the robot in a square pattern.

```
rLocate 300,500
for n = 1 to 4
  rForward 120
  rTurn 90
next
```
Figure 31: A loop is used to repeat the movement commands four times, causing the robot to move in square.

25

Leaving a Trail

We can cause the robot to drop a pen and draw a line as it moves by adding the statements shown in Figure 32. The **rInvisible** command tells the robot to ignore objects of color **GREEN**. Without this command, the robot will see the line it is drawing as an object in the room. This will cause an error because the robot will think it has collided with an object. The **LineWidth** statement forces the line being drawn to be four pixels wide, making it easier to see. The **rPen DOWN** statement lowers pen, which automatically draws using the first color in the invisible list.

```
rLocate 300,500
rInvisible GREEN
LineWidth 4
rPen DOWN
for n = 1 to 4
  rForward 120
  rTurn 90
next
```

Figure 32: The program in Figure 1.6 can be modified so that the robot draws a line as it moves.

When the program in Figure 32 is run, it produces the screen shows in Figure 33. Notice that repeating the movement commands four times does move the robot in a square motion and returns it to its original starting position.

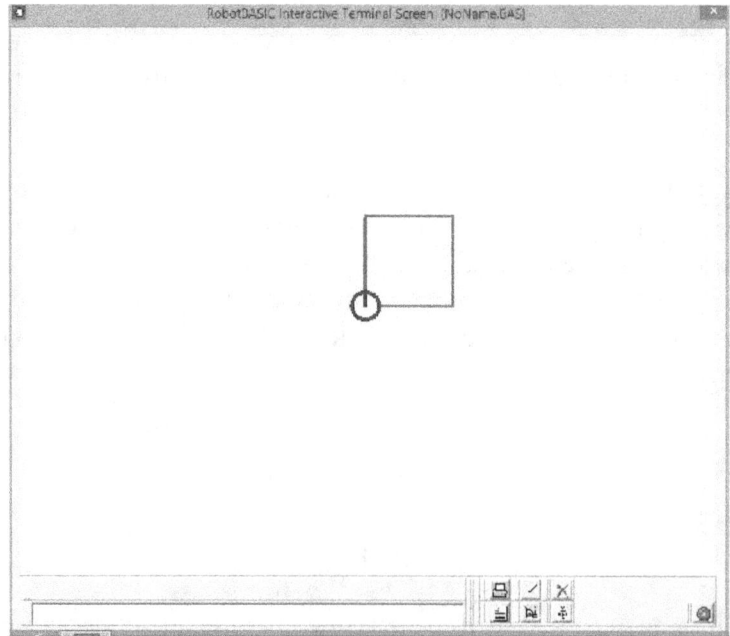

Figure 33: This screen is created by the program in Figure 32.

Organizing Your Program

The program in Figure 32 is still small, but it is never too early to think about organizing your programs so they are easier to read and understand as they get bigger. This can be done by placing functional pieces of your program in separate modules called *subroutines*. The subroutines can be executed with a **gosub** statement when needed, from the main program as shown in Figure 34. Notice that subroutine names must end with a colon when you are defining a new subroutine. Subroutine names, just like variables, are case sensitive. When a subroutine terminates by executing a **return** statement, the program continues with the line following the **gosub** that originally directed flow to the subroutine.

Notice that the program now is made up of two main modules, one that initializes the robot and one that causes it to draw a square. Notice also that the names of the modules reflect their function. This not only makes the program easier to understand, it creates modules that are easier to re-used when needed. Notice the **Main** program simply executes each subroutine by calling them with a **gosub** statement. The name **Main** just helps you visualize where the program starts. You could use any name though, or even no name at all, as RobotBASIC programs start at the first executable statement found in the file.

```
Main:
  gosub InitializeRobot
  gosub DrawSquare
end

InitializeSimulator:
  rLocate 300,500
  rInvisible GREEN
  LineWidth 4
  rPen DOWN
return

DrawSquare:
  for n = 1 to 4
    rForward 120
    rTurn 90
  next
return
```
Figure 34: This program is a more organized version of the program in Figure 32.

Notice also that each module in Figure 34 is indented to make it easy to identify where the module begins and ends. Notice also that this indenting is also used with loops to help see where the start and stop. Indenting is not required, but it will make your programs far easier to read, especially as they get larger.

Real World Considerations

Recall from Figure 33, that the robot moved in a square pattern. In the real world a robot is not likely to have the precise movement demonstrated by the Figure. It might, for example, turn slightly more or less than 90° or move slightly more or less than the requested distance. This can happen, for example, because one motor is slightly faster than the other – perhaps because it has better bearings and thus less friction (although there are many reasons for this type of inaccuracy).

One way to make a real robot's movements more precise is have encoders on each motor that count pulses that are generated by a special circuit as the wheel moves. These pulses allow the robot's computer to keep track of how far and how fast each motor moves. Advanced programs can use this information to dynamically change the speed of the wheels so they each move the same amount, at least as accurate as the amount of movement associated with each encoder pulse. If for example, a wheel produces 100 pulses per revolution, then the computer should be able to know the wheel's position within 1%. This limitation, plus the fact that one wheel might slip slightly on the floor, means that a real robot will always have some error associated with its movement.

We can force the simulated robot to generate some percentage of random error by altering the **InitializeSimulator** module as shown in Figure 35. Notice also, the comment explaining the purpose of the line. Any text following double slashes is considered a comment and is ignored by RobotBASIC. If you make this modification and run the program again, the movement will look something like that shown in Figure 36. Remember, the actual movement will be different every time the program is run, because the error generated is random (just like the movement of a real robot).

```
InitializeSimulator:
  rLocate 300,500
  rInvisible GREEN
  LineWidth 4
  rPen DOWN
  rSlip 15 // produces up to 15% random error
return
```

Figure 35: The rSlip command tells the simulated robot to create random error.

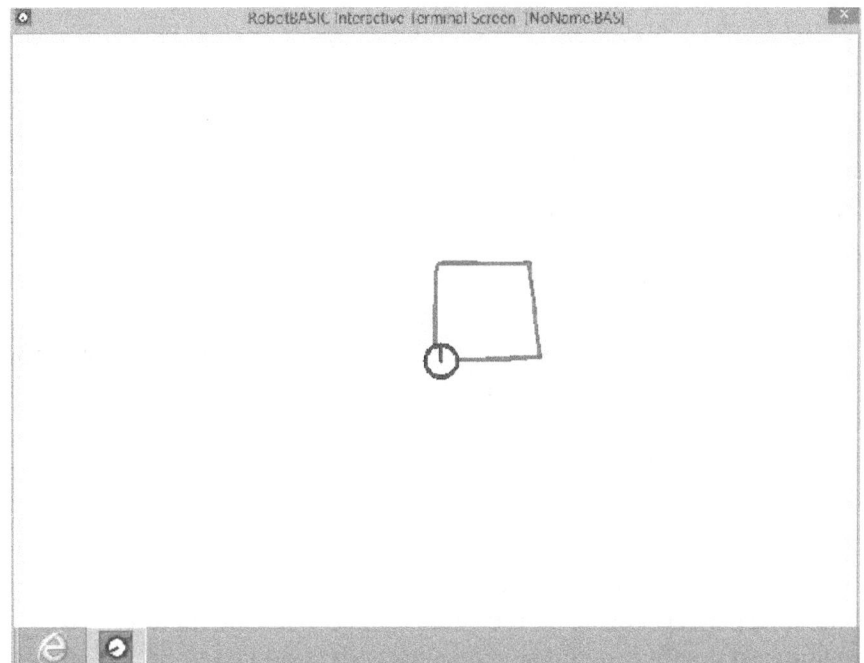

Figure 36: Now the simulated robot performs much more like a real robot.

If you study more about the RobotBASIC simulated robot, you can learn how the robot can sense objects in the room and use that information to constantly correct its motion in order to achieve goals that you create for it, even if its movements are not precise. For example, a robot with an electronic compass might compare the direction it thinks it is facing to the compass reading and then make appropriate corrections. When sensors are used in this manner, the robot's normal movements do not have to be precise because errors are constantly corrected based on the sensory information.

Controlling a Real Robot
You can actually use RobotBASIC programs that control the simulator to control real robots. Visit www.RobotBASIC.org for more information.

Back to the Racetrack Timer
Now that you know a little about programming let's learn how to create a program that can time your car. RobotBASIC has a command that can tell you what mouse buttons are pressed (and where the mouse pointer is on the screen). The command is:

```
ReadMouse x,y,b
```

When this line is executed the variables **x** and **y** will contain the location of where the mouse pointer is on the screen. The value of the variable **b** will tell us which, if any, of the buttons is pressed. You can use any variable names in your programs. The names **x**, **y**, and **b** are just examples. Look at the program in Figure 37.

29

```
repeat
  ReadMouse x,y,b
  Print x;y;b
until false
```
Figure 37: This program shows how to display mouse information.

If you type in the program in Figure 37 and run it, it will continually print three things (the x and y coordinates of the mouse, and a number indicating which button is pressed). The reason the printing is continuous is that the two main statements in the program are inside a loop. The **REPEAT** statement starts the loop and in this case it continues forever because it continues until the expression following the **UNTIL** statement is true (and in this example, it is always false). More on this later. The **UNTIL** statement marks the end of the loop.

After you run the program, move the mouse around and watch the values change for the coordinates. The upper-left corner of the screen is 0,0 with the numbers getting bigger going to the right and down. Since the loop causes the two statements to execute over and over, the program continually obtains information about the mouse and prints it. The first print will be at the top of the screen with each new print on the next line. When the last line on the screen is reached, the screen clears automatically and new printing starts again at the top. This continuous printing make the output hard to read, but we will fix that shortly. For now, you might add a delay as shown in Figure 38. It makes the program delay 100 milliseconds (1/10 of a second) between each read and print action.

```
repeat
  ReadMouse x,y,b
  Print x;y;b
  delay 100
until false
```
Figure 38: The delay makes the output more readable.

Notice also that the third number printed on each line is zero. This simply means that no mouse buttons have been pressed. If you press the mouse buttons while the program is running you will see that a one is printed when the LEFT button is pressed and a two is printed when the RIGHT button is pressed. Note: The mouse cursor has to be in the terminal window for this to work. You can test this program using your regular computer mouse or the modified mouse. The remote switches should change what is printed (as described above) if you have modified your mouse correctly.

If you run this program you might be annoyed at the constant printing down the screen. The **PRINT** statement we are using is very easy to use, but it does have its limitations. RobotBASIC has another PRINT-type command that has more options. It is called

xyString. Figure 39 shows how **xyString** can be used to make the programs in Figures 37 and 38 perform better.

The first two numbers in the **xyString** command determine where the information will be printed. In this example, the starting location is 200 pixels across and 100 pixels down. The remaining parameters in the command are exactly like the **PRINT** statement. Notice that the **delay** is no longer necessary.

```
repeat
  ReadMouse x,y,b
  xyString 200,100, x;y;b
until false
```
Figure 39: The delay makes the output more readable.

It was mentioned earlier that the **REPEAT** loop repeats all the statements in it as long as the express following the UNTIL is true. In our example, it was always true, so it never ended (unless you terminated the program manually). The program in Figure 40 terminates when you press the RIGHT mouse button because it continues until the variable **b** has a value of 2 (which happens when you press the RIGHT mouse button).

```
repeat
  ReadMouse x,y,b
  xyString 200,100, x;y;b
until b=2
```
Figure 40: The delay makes the output more readable.

The Timing Program
These simple examples are the basis for building a timing program. Figure 41 shows a complete program that can time your car. Remember, you can obtain more information on any command you are not familiar with by using RobotBASIC's HELP file.

The program is full of comments to help you understand how it works, but there are a few things that are worth discussing. The main part of the program is enclosed in a **WHILE-WEND** loop. It is much like a **REPEAT-UNTIL** loop except a **WHILE** loop decides whether to end the loop at the beginning rather than the end of the loop. The **WHILE** loop continues looping when the expression following the **WHILE** is true. This is opposite of the **REPEAT-UNTIL** loop, which continues looping while the expression is false (it stops when the expression is true). In this example, the **WHILE** loop keeps the programming running until it is manually stopped.

Inside the **WHILE**, are several **REPEAT** loops that watch the mouse and wait for certain conditions to occur (as depicted by the comments in the program). The program assumes the LEFT mouse switch is at the car's starting position and the RIGHT mouse switch is at the lower end of the ramp.

RobotBASIC has a **Timer()** function that gives the current time (in milliseconds) of an internal counter. The time of the car's trip can be found by subtracting the original timer value from the current timer value. For this reason, the program should store the current time when the left mouse button is released (indicating the car has left its starting position).

```
xyString 200,100,"Place car on starting position.        "
while true
  repeat    // wait for the car to be placed at start
    ReadMouse x,y,b
  until b=1
  xyString 200,100, "Car moved to starting position.      "
  xyString 200,200, "Waiting for car to be released.      "
  repeat    // wait for car to be released
    ReadMouse x,y,b
  until b=0
  xyString 200,100, "Timer has been started.              "
  // start timer by recording current time
  t=timer()
  // wait till car reaches end of track
  // or car is placed back on the starting position
  repeat
    ReadMouse x,y,b
    // calculate the current time
    cur = (timer()-t)/1000
    xyString 200,200,cur," seconds.                       "
  until b>0
  // print appropriate messages depending on mouse button
  // terminated the above loop
  if b=1
    // do this if left button was pushed
    xyString 200,100, "Car has been moved to start.        "
    // clear timer display since car moved back to start
    xyString 200,200, "                                    "
  else
    // do this if right button was pushed
    xyString 200,100, "Place car on starting position.     "
  endif
wend
```

Figure 41: Complete car-timing program.

One thing that complicates the program slightly is that it handles unexpected situations such as the car getting stuck or jumping off the track. If things like this happen we don't want the user to have to trigger the final switch to get the program back in sync. Such situations are handled by the final **REPEAT** loop which terminates if *either* mouse button is pressed. This allows the program to start over automatically when the car reaches the second switch or when the car is simply replaced at the starting position.

There are **xyString** statements throughout the program to keep the user informed by printing messages explaining what is going on. At the very end of the program, an **IF-ELSE-ENDIF** control structure prints different things depending on whether it was the RIGHT or LEFT button that terminated the final **REPEAT** loop.

After you have some experience with RobotBASIC you may want to modify this program with special features of your own. For example, if you use **xyText** instead of **xyString**, you will be able to increase the size of the font to make the messages and time easier to read from a distance (spectators might enjoy this).

You could also use color for some of the messages or even add graphics to show where the car is on a simulated ramp. Your choices are really boundless. Just study RobotBASIC's integrated HELP file or one of our low-cost books to learn more about RobotBASIC programming.

Enhancements

The use of switches to detect when your car starts and stops is, by far, the easiest and cheapest way to build this project. There are other options, though. For example, you could use an IR (infrared) sensor (instead of the switches) to detect when the car starts and when it reaches the end of the track. Some people might prefer this because nothing has to touch the car. Simply mount one detector so it detects the car when it is at the starting line and a second detector at the end of the track to detect the car as it rolls past.

The good news is that the infrared option works just like switches as far as the computer is concerned. This means that the program in Figure 41 will still work without modification.

While it is possible to create your own circuits for generating and detecting the light beam, it is much easier (and cheaper) to purchase a sensor than was designed for the purpose. Figure 42 shows such a sensor that can be purchased from www.Pololu.com (Item # 1134). You can solder wires directly to it or plug it into a breadboard such as the one in Figure 43 (also available from Pololu). The sensor emits invisible IR light and detects when it is reflected back from nearby objects.

Figure 42: This GP2Y0D810Z0F IR sensor can detect objects up to four inches away.

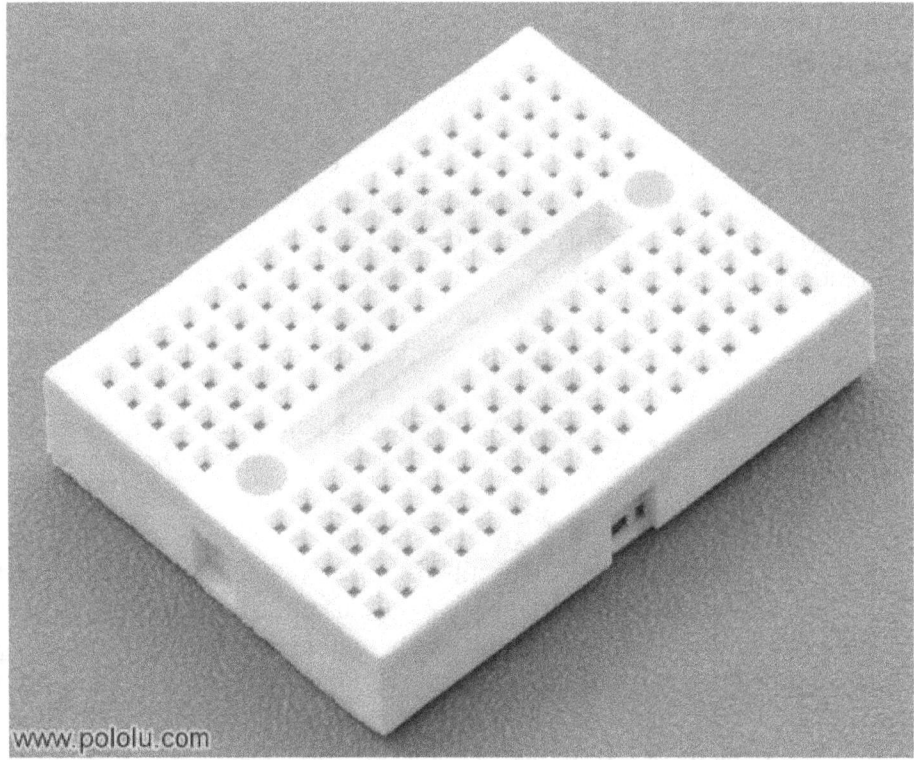

Figure 43: Breadboards such as this make it easy to connect components together.

The IR Sensor in Figure 42 is less than one inch tall and can detect objects up to four inches away. Even better, when properly connected, this sensor can replace the remote switches in our previous diagrams. Figure 44 shows the circuit necessary to do this.

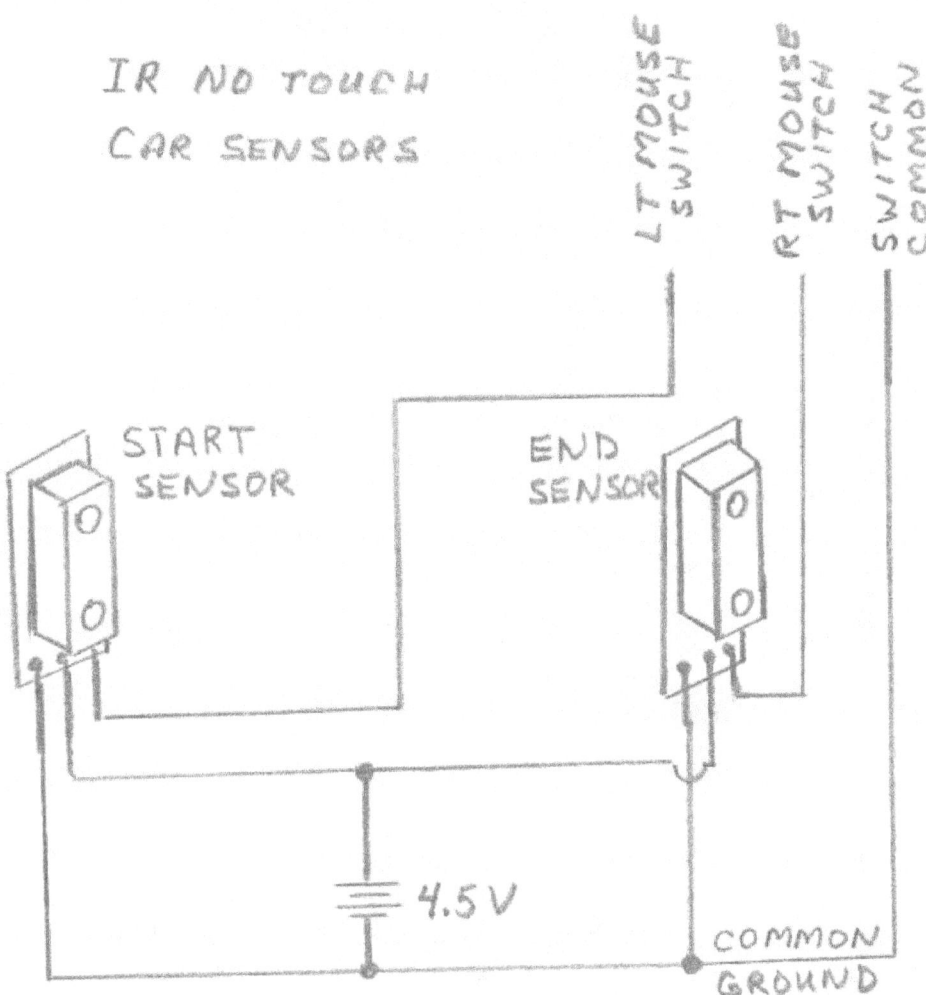

Figure 44: IR sensors can replace the car timer's remote switches.

The IR sensors must be powered by 4.5 to 5 volts. This is easy to do with three batteries using a holder like the one in Figure 45. The black wire on the holder is the negative lead and should be connected to the common terminals. Battery holders are available from companies like Pololu and Radio Shack.

www.pololu.com

Figure 45: A battery holder makes it easy to obtain the power for the IR sensors.

Skill and Knowledge

It must be stressed that building a timer using the IR sensor approach requires more skill and knowledge than using switches – which is why it is suggested that most people opt for the switch solution. If you are capable or if you have a friend or teacher with some electronic experience, then the IR approach can produce a more sophisticated system. Don't think though, that you need to use IR. The low-tech switch approach can perform with equal results. The IR alternative is only offered for those looking for a more high tech solution.

Final Word

This is a relatively long document for such a small project. I could have just told you *what* to do without taking the time to explain how things work. In the end, I think most scouts will enjoy learning a little about electricity and programming. With a little thought, you might even use some of your newfound knowledge to obtain a Merit Badge for a related area.